My Birthday Party

Dona Herweck Rice

Publishing Credits

Rachelle Cracchiolo, M.S.Ed., *Publisher*
Conni Medina, M.A.Ed., *Managing Editor*
Nika Fabienke, Ed.D., *Content Director*
Véronique Bos, *Creative Director*
Shaun N. Bernadou, *Art Director*
Carol Huey-Gatewood, M.A.Ed., *Editor*
Valerie Morales, *Associate Editor*
Courtney Roberson, *Senior Graphic Designer*

Image Credits: All images from iStock and/or Shutterstock.

Teacher Created Materials
5301 Oceanus Drive
Huntington Beach, CA 92649-1030
www.tcmpub.com

ISBN 978-1-4938-9827-5

The  is from  .

balloon grandma

The  is from  .

hat grandpa

The  is from  .

cake mom

The  is from  .

ice cream dad

The  is from .

doll grandma

The  is from  .

bike grandpa

The  is from  .

ball mom

The  is from  .

bear dad

The  is from  .

hug me

The  is from  .

hug me

High-Frequency Words

from

is

the